1

PRACTICAL GUIDE TO PRODUCTION PLANNING & CONTROL
[REVISED EDITION]

KERWIN MATHEW

PREFACE

This book covers production planning and control concepts such as MRP and MRP II and acts as a practical guide for all personnel directly or indirectly involved in production management and/or production planning and control, e.g., production planners and engineers.

For students of production engineering or management subjects, it will act as an easy-to-understand guidebook.

Computer personnel involved with computer-aided manufacturing will also find the book useful.

The book avoids technical jargon as far as possible in its attempt to explain the functions and work of the production planning and control department, the nerve center of any manufacturing organization.

Kerwin Mathew, Ph.D., PE, CMfgT, CPM

CONTENTS

1 INTRODUCTION TO PRODUCTION PLANNING AND CONTROL

Production comprises of a sequence of operations which transform materials from a given to a desired form. The greatest efficiency in production is achieved by manufacturing the required quantity and quality of the product at the right time and at the lowest possible costs. For attaining this, production planning and control is one of the important tools. It enables the direction and coordination of the firm's material and physical facilities towards achieving the respective objectives in the most efficient way.

In the implementation of production planning and control, two important goals have to be borne in mind and attained, namely, producing the required quantities of a given product and producing these quantities at appropriate times. It should be noted that the system of planning and control adopted and the procedures involved will vary from organization to organization, for instance, that in a firm with data-processing equipment will differ somewhat from that in a firm which does not possess such equipment.

However, regardless of how elaborately a system is developed, things will not work out perfectly at all times. Operation schedules which are developed by the most scientific methods will be disrupted when an employee decides to take the day off or when a supplier makes an incorrect shipment or when someone inadvertently jams his tool into the working parts of a machine. Planning and control activities furthermore involve the use of judgment or are based on predictions of the future course of events, and errors in judgment and in forecasting are common-place.

Also, in some companies, production planning and control is undertaken by the production control department, while in others, the function is part of the job of the manufacturing department. However, whether it is necessary to have a production control department or not does not matter, as there are organizations which produce goods efficiently despite the absence of a production control department.

Definition of Production Planning and Control
Production planning and control is a staff function which ensures that targeted outputs based on sales orders or forecasts are produced on time, in the right quantity and quality, and at the lowest possible cost.

Objectives
The goals of production planning and control could be as follows:-

(a) To determine when a job is to be carried out.
(b) To decide where the work is to be carried out in cases where there are similar machines or

work-places which could be used as alternatives.

(c) To ensure that materials or purchased parts are ordered at the right time to fit in with the plan of work scheduled for the factory.

(d) To ensure that the proper person is performing the proper work in the specified time, consuming specified materials, and at the specified quality.

Control is needed to ensure that:

(i) the plan is implemented

(ii) adjustments are made when the plan fails or when external circumstances warrant it

Functions

Production planning and control could be considered the nerve center of the factory. A completely equipped factory might be ready to function, but without production control not one move could be made, not one wheel could be turned.

A continuous stream of directions has to flow to the factory so that there would be continuing, coordinated activity and so that products are manufactured in the desired form, quantity and quality, and, on time and within the limits of the cost permitted.

Production control normally refers to the internal day-to-day direction of factory operations. Orders from the customer or the sales department and raw materials are the two main input factors in production. Both orders and materials flow through the plant; it is the production control function which directs this flow.

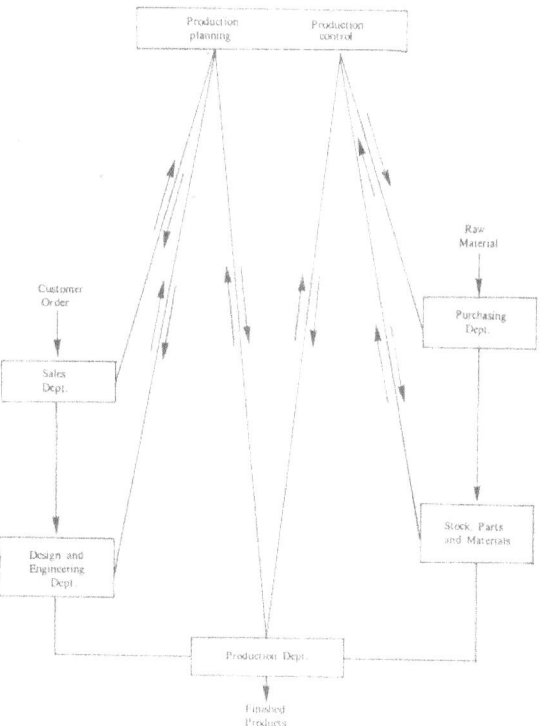

Fig. 1 Indicates how the production control function is influencing
all main activities in the company.

The work of production control could be carried out in many varied ways, but there are rather few variations in the fundamental functions of that work. Regardless of the nature of the industry or the size of the company, these functions apply to all manufacturing plants.

The following are the major functions of production control:-

(1) Receiving orders for products from the sales department.
(2) Determining the finished parts required.
(3) Determining the materials required.
(4) Maintaining the stock of raw materials.
(5) Determining the required operations.
(6) Determining the machines and machine attachments needed.
(7) Determining the sequence of operations.
(8) Putting up production orders.
(9) Putting up production schedules.
(10) Ensuring that all production facilities are available.
(11) Distribution of jobs to men and machines.
(12) Giving instructions to produce (despatching).
(13) Controlling the transportation of materials.
(14) Receiving reports on progress and evaluating performance.
(15) Re-planning when original plans are not executed.
(16) Initiating remedies when original plans are not executed.
(17) Controlling the stock of finished parts.
(18) Controlling the stock of finished products.
(19) Attending to requests for information from the sales department.
(20) Estimation of costs of new jobs.

Whether assigned to a production planning and control department or not, the above functions are basic and have to be performed by some departments in the organization.

High volume repetitive work would simplify the performance of these functions while varied work would complicate it.
The place, the frequency and how detailed the functions are being performed would vary. The functions have to be carried out some time, somehow by someone, irrespective of the variations.

The following are functions which may be associated with production control:-

(a) Checking on shipping requirements (containers, et al.).
(b) Instituting the utilization of pallets or other materials handling equipment.
(c) Operating the mail distribution service.
(d) Standardization of parts and materials.

Though the other functions are important for the proper execution of the 20 basic functions listed above, they are not essential. Excluding the store functions, these 20 basic functions could be broadly divided and classified into the following four main functions:-

(a) *Preplanning* - estimating cost, advanced planning of the special sequence of operation and checking the availability of resources.
(b) *Scheduling* - plotting a detailed time-table for operations.
(c) *Despatching* - executing the scheduling and routing of materials.
(d) *Following up and controlling* - controlling and making the necessary adjustments after the program has been executed.

Type of Production

The principles of production planning and control are general, i.e., they apply all plants or factories, products, types of production and types of companies.

The methods of application would however vary and would depend mainly on the following factors:-

(1) Whether the products are standard or special purpose.
(2) Whether the products are manufactured according to customer orders or in anticipation of demand as stock.
(3) Whether production is single project, intermitten or continuous.

As is shown in Fig. 2, these factors are more or less inter-related. It may be said that the type of products and the type of orders determine the type of production.

Job or Single Project Production

In the single project type of production, special products are manufactured in small quantities, with no order expected to be repeated, e.g., shipbuilding, special transformers, special transportation and handling equipment, turbines, et al.

Though there is the typical non-uniformity of products being manufactured in this type of production, there are certain common characteristics which have some bearing on the planning and control methods to be employed.

Batch or Intermitten Production

In intermitten production, a limited number of rather standardized products are manufactured on a lot-basis, with lots repeated somewhat regularly. This requires that the manufacturing process be divided into sub-processes, and that each sub-process is completed for the whole batch before the next sub-process is undertaken, e.g., consumer goods produced for a limited market such as electronic instruments, household appliances, production equipment such as machine tools and hand tools, et al.

Intermitten production essentially differs from single project production in that the number of products are being made repeatedly.

Flow or Continuous Production

In continuous production, the materials move continuously, in a uniform rate, through a sequence of balanced operations towards completion, e.g., in automobile assembly.

The important characteristics of continuous production are large volume and small variety of products made.

For there to be continuous production, there has to be balanced capacity of successive machines.

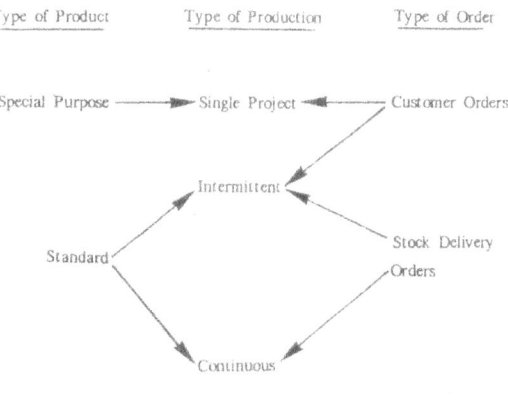

Fig. 2

The different characteristics of the three types of production will be discussed in some detail.

Questions for Review
(a) Give the responsibilities of the production planning and control department.
(b) Who establishes the shipping dates for customers and ensures that they are met?
(c) How is the type of production determined by the type of product and the type of order?

2 FORECASTING

Careful and accurate planning of inventories and production is dependent on the reliability of sales forecasts. The sales forecasts are usually prepared by the sales or marketing department, but they are sometimes prepared as a team effort by several departments or by another appropriate department.

The minimum amount of forecasting required for providing an adequate basis for inventory planning is an annual forecast of the expected sales of finished products, which should be firmed up every quarterly. The quarterly figures are utilized for commitments, e.g., the purchase of materials, and scheduling manufacturing. The annual ordering quantities, the expected commitments with suppliers, and the man-hours and machine-hours required to meet the anticipated production volume are affected by the annual forecast.

The quarterly trend should be a continuous three months' projection, with each month adding one and dropping another, so that action could be taken promptly as required to meet the needs of changing trends as each new month is added.

Preparing the Sales Forecast
As a basis for inventory planning and production planning and control, sales forecasting is often concerned with the anticipated sales of a single product or a product classification during a given period, such as a year or a stated number of months. To be useful, the sales forecast must either be presented in terms of production units or be readily convertible.

Forecasting on the Basis of Past Performance
Many companies base their sales forecasts on past performance factored, as needed, to reflect anticipated changes in the economy or other important factors. The sales forecast for the coming year or month is based on the number of units sold last year or for the same month last year. This historical forecast is very useful when seasonal variations play a major role. A soft drink manufacturer, e.g., can predict with reasonable confidence the expected peak sales periods of the year and plan to produce for stock in the low sales period.

It is helpful to plot the information in a way to disclose a sales trend when reliable historical information is available for a period of several years. A simple trendline diagram is shown in Fig. 3. On the vertical axis units of sales activity are plotted while on the horizontal axis time is plotted. To indicate the probable trend, a single line is fitted to the line representing actual sales.

It is often feasible to utilize more complex statistical methods for analysing historical sales information for sales forecasting purposes, such as regression analysis and time series analysis, with the help of modern data processing equipment.

Forecasting on the Basis of Sales Force Expectations

Sales forecasts are frequently based on the premise that the salesmen and others in the sales division are best able to estimate the pattern of future sales activity.

Each salesman first estimates the volume of his probable sales on the basis of specific prospects and the product requirements of each territory, a technique which is often referred to as "customer-prospect audit". The branch managers and district managers, as well as the successive levels of sales management, would review these preliminary forecasts, or audits, taking advantage of the specialized knowledge and judgments of each level.

The sales forecast is finally studied from the standpoint of whether it could be met from the manufacturing point of view.

Manufacturing capacities, material availabilities, manpower requirements, costs and other related factors must be given full consideration.

Developing a forecast on the basis of sales force estimation is most effective when only a few products are involved and when relatively few customers account for a large portion of the sales.

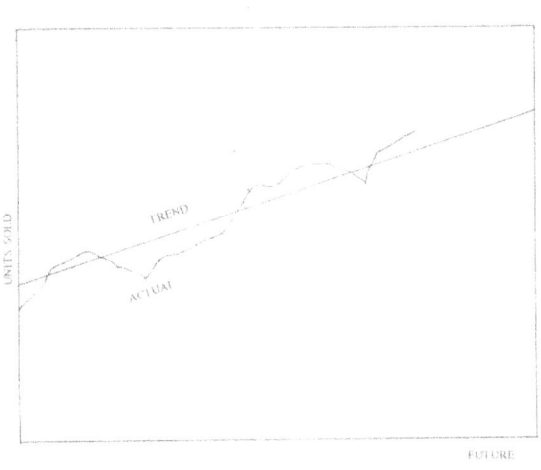

Fig. 3 A simple Trendline Diagram

Questions for Review

(a) How do you enhance the effectiveness in forecasting when the available data for forecasting is never completely accurate and applicable?

(b) Is it necessary to forecast the sub-assemblies and components required after completing the forecasting of finished goods? How are their requirements determined?

3 PREPLANNING

Preplanning concerns the action taken before production. The sales department will inform management that a product can be sold in the market at a certain price with a certain demand. This information is provided to the production planning and control department for the estimation of costs and for the determination of how the product is to be made.

Cost Estimation
Cost estimation in production planning concerns the determination of production costs before actual production.

Accurate estimates are necessary for the success of all companies. Estimates serve the following purposes:-

(a) To submit price quotes to customers before obtaining orders.
(b) To evaluate future costs of existing products wherein changes in sales volume are planned or expected.
(c) To determine the selling prices of new products or products with changes in design.

Generally, cost estimates are more important to companies producing to customers' orders than to companies producing in anticipation of demand. Product designs and quality standards are established in accordance with the expected selling price. If production costs exceed the estimated costs, serious losses may result.

Technical Planning
When the order is given to manufacture, the production planning and control department works out the details of all the manufacturing processes. In carrying out this task, he could utilize a process sheet (shown in Fig. 4) and also determine what parts are to be obtained from outside sources.

Basically, a process sheet is an outline of the necessary manufacturing operations. The process sheet will also indicate the machines, methods and tools to be used and the standard time to be taken foe each step in the process.

The production planning and control department makes great use of the data of the time required to accomplish each operation provided in the process sheet as a basis for establishing delivery dates, machine loading and scheduling.

Resource Checking

Checking the availability of resources, namely, men, machines and materials, is the final stage in preplanning. It is important that prior to accepting an order the following are checked:-

(a) The credit status of the customer.
(b) The current loading of machine and manpower.
(c) The raw materials' availability.
(d) The availability of the tools which are required for manufacturing the product.

PROCESS SHEET

Description (or part): _____ Part No. _____

Material Specification: _____ Size: _____

Quantity Required: _____ Unit: _____

Scrap Allowed: _____ Drawing No. _____

Operation No (1)	Operation Description (2)	Department (3)	M/C Code (4)	Tools Used (5)	Standard Time (6)	Remarks
1	Mould	02	01	T-100	1.2 hrs/n	

Issued by _____

Date _____

Fig 4 Process sheet

Questions for Review
(a) What is the objective of cost estimation?
(b) How do you carry out technical planning by using the process sheet?
(c) In the final stage of preplanning, what have to be checked before an order is accepted?

4 PRODUCTION SCHEDULING

Scheduling establishes the time on when the work is to be carried out. It can be defined as the process of fitting jobs into a logical timetable. Good scheduling is vitally essential for any real efficiency in the manufacturing program as a whole, regardless of the type of industry.

A smooth, timely flow of products or components through the various manufacturing operations should be achieved by the planning function of production scheduling. Commencing from the requirements of customer orders or from the sales/inventory forecasting systems, planning covers the loading of work orders onto the shopfloor departments in such a way that parts are made available for downstream operations such as sub-assemblies and final assembly or the completed products.

The following are the three broad categories which production scheduling can be divided into:-

(a) Master Schedule
(b) Parts Schedule/MRP
(c) Machine Loading Schedule

Master Schedule
The master schedule covers all the products made in the factory. It keeps the various assembly departments informed about the items they have to produce within each period, e.g., every week or month. A specimen master schedule for a fountain pen project is provided below:-

Master Schedule For Manufacture Of Fountain Pens
Total production per year - 960,000

		Yearly	Monthly
Black	....	480,000	40,000
Red	....	240,000	20,000
Green	....	120,000	10,000
Blue	....	120,000	10,000
		960,000	80,000

The master schedule can be prepared in a graphical way, i.e., as a Gantt Chart. In the Gantt Chart the monthly production can be indicated as follows:-

Manufacture of Fountain Pens

S/No	Description	Jan	Feb	Mar	Apr	May
1	Black	40,000	40,000	40,000	40,000	40,000
2	Red	20,000	20,000	20,000	20,000	20,000
3	Green	10,000	10,000	10,000	10,000	10,000
4	Blue	10,000	10,000	10,000	10,000	10,000
	Total	80,000	80,000	80,000	80,000	80,000

In the above example, constant monthly production targets are established for the year. The rates of output may be altered if necessary to be more in line with the management's forecasts of market demand. If, e.g., the forecasts indicate that sales would be slow in the beginning of the year but would increase in April and May, the master schedule might be modified as follows:-

S/No	Description	Jan	Feb	Mar	Apr	May
1	Black	28,000	30,000	40,000	50,000	50,000
2	Red	12,000	15,000	20,000	25,000	30,000
3	Green	5,000	10,000	10,000	15,000	20,000
4	Blue	5,000	5,000	10,000	10,000	10,000
	Total	50,000	60,000	80,000	100,00	110,000

Horizontal lines drawn to scale could be incorporated below the monthly figures in order to achieve greater visibility and improve control as production progresses. We could indicate the target by, e.g., a thin line and the achievement or result by a thick line. As we proceed, we could draw a cumulative thick line giving the cumulative production.

For it to be easily read, the chart could be slightly modified, e.g., a color scheme could be used, yellow representing quantity scheduled, red super-imposed indicating quantity accomplished and blue showing cumulative total. A yellow portion visible in the chart means that production is behind schedule. In using this chart, the principle of management by exception is adhered to. The manager should take action only when a yellow line is seen. Non-accomplishment of the production target could be indicated in the chart by using symbols (refer to Fig. 5), e.g.:-

A = Lack of raw material
B = No operator
C = Power failure
D = Unexpected plant break-down

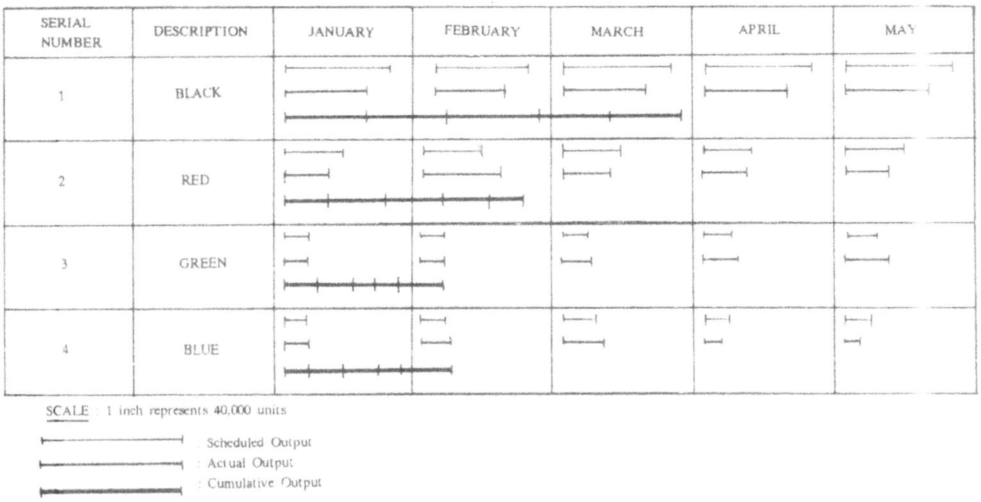

SERIAL NUMBER	DESCRIPTION	JANUARY	FEBRUARY	MARCH	APRIL	MAY
1	BLACK					
2	RED					
3	GREEN					
4	BLUE					

SCALE : 1 inch represents 40,000 units

⊢――――――⊣ : Scheduled Output
⊢――――――⊣ : Actual Output
⊢――――――⊣ : Cumulative Output

Fig 5: GANTT CHART SHOWING MASTER SCHEDULE, ACTUAL MONTHLY OUTPUTS AND CUMULATIVE OUTPUTS OF FOUNTAIN PENS

The chart condenses all information and conveys it simply and plainly.

For more effective control, it is advisable to have a weekly schedule. As the number of holidays in a year will add up to two weeks, it is better to schedule for 50 weeks in a year. Any balance left over or unaccomplished in the 50-week schedule due to the interference of holidays could be easily covered during the last two weeks.

Where it is necessary to increase production at any particular stage, it should not be carried out hastily. Production should instead be increased in stages so that the changes would not be too sudden, the department concerned would have no problem coping and adjustments could be more easily made.

Scheduling should be carried out only after studying the work contents of the job and the factory capacity. Changes are sometimes made without considering the products' work contents. This could lead to problems. Therefore, in preparing a schedule attention should be given to the following:-

(a) The capacities of the various sections or departments concerned.
(b) The efficiencies of the various sections or departments concerned.
(c) The maintenance schedule.
(d) Holidays.
(e) Anticipated sickness/absenteeism.
(f) Availability of materials.
(g) Existing commitments.

Parts Schedule/Material Requirements Planning (MRP)
A Material Requirements Planning system generates a complete list of parts and sub-assemblies needed to produce the end item along with the required quantities and the correct timing for releasing orders for these items, based on the demand for the end item. That is, Material Requirements Planning creates schedules identifying the specific parts and materials needed to produce the end item, the exact quantities needed, and the dates when orders for these materials should be released and be received or completed within the production cycle. The term Material Requirements Planning, or MRP for short, implies the utilization of a large computer program (usually one produced by a major computer manufacturer) to carry out the foregoing operations.

Material Requirements Planning is not a new concept. The Romans had probably used it in their construction projects, the Venetians in their shipbuilding, and the Chinese in constructing the Great Wall. Due to space constraints, building contractors have always been forced into planning for materials to be delivered when required and not before. The use of high capacity computers in Material Requirements Planning allows for the larger scale of work carried out and more rapid changes, enabling the firms utilizing Material Requirements Planning to produce many products involving thousands of parts and materials.

The objectives of inventory management in the Material Requirements Planning system are to improve customer service, minimize inventory investment and maximize production operating efficiency.

Material Requirements Planning sees to it that materials are expedited (hurried) when their lack would delay the overall production schedule and de-expedited (delayed) when the schedule lags behind.

Though the term Material Requirements Planning (MRP) refers to a computer program which generates schedules to meet material requirements, it is also used to imply the total system of materials planning, which also includes the inputs to the computer program, as is shown in Fig. 6. The MRP system essentially works as follows: Based on orders for the products, a Master Production Schedule which states the quantity of items to be manufactured during specific time periods is created. A Bill of Material File which identifies the specific materials that are to be used to produce each item and specifies the correct quantities of each is also created, as well as the Inventory Records File which contains data such as the number of units on hand and on order. The Master Production Schedule, Bill of Materials File and Inventory Records File make up the data sources for the Material Requirements Program, which breaks down the production schedule into a detailed order schedule plan for the entire production sequence.

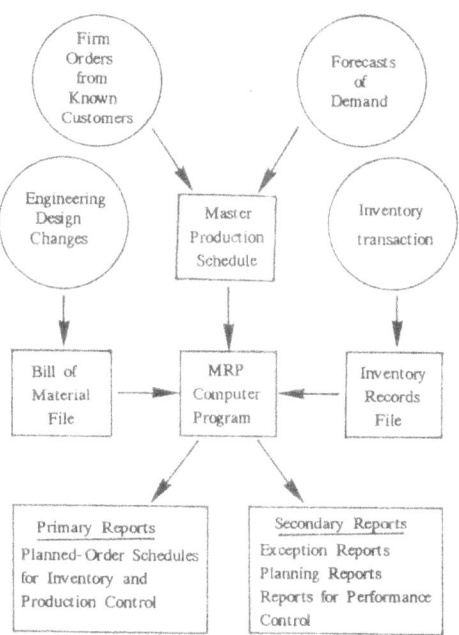

Fig 6

N.B. Overall view of the inputs to a Material Requirements Planning
program and the reports generated by the program.

Example

Product K has the product structure shown below. Complete the MRP tables if the lots sizes are equal to one.

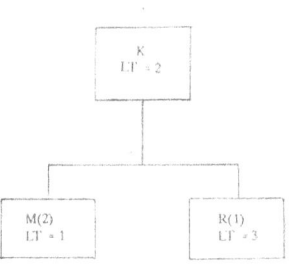

					Period			
PD	1	2	3	4	5	6	7	8

K

Gross Requirements		25	15	120	0	60	0	15	0
Scheduled Receipt								
Projected On Hand	50							
Net Requirements								
Planned Order Receipts								
Planned Order Releases								

M

Gross Requirements								
Scheduled Receipts		30						
Projected On Hand	225							
Net Requirements								
Planned Order Receipts								
Planned Order Releases								

ITEM: M   LEVEL CODE: 1   LEAD TIMES: 1   LOT SIZE: 1   SAFETY STOCK: 0

ON HAND INVENTORY	225																			
ALLOCATED		PERIOD	1	2	3	4	5	6	7	8	9	10	11	12	13	14	15	16	17	18
GROSS REQUIREMENTS			220	0	120	0	30	0												
SCHEDULED RECEIPTS			30																	
AVAILABLE INVENTORY	225		35	35	0	0	0	0												
NET REQUIREMENTS					85		30													
PLANNED ORDER RECEIPTS					85		30													
PLANNED ORDER RELEASES				85		30														
PLANNED AVAILABLE INVENTORY																				

ITEM: K   LEVEL CODE: 0   LEAD TIME: 2   LOT SIZE: 1   SAFETY STOCK: 0

ON HAND INVENTORY	50																			
ALLOCATED		PERIOD	1	2	3	4	5	6	7	8	9	10	11	12	13	14	15	16	17	18
GROSS REQUIREMENTS			25	15	120	0	60	0	15	0										
SCHEDULED RECEIPTS																				
AVAILABLE INVENTORY	50		25	10	0	0	0	0	0	0										
NET REQUIREMENTS					110		60	0	15	0										
PLANNED ORDER RECEIPTS					110		60		15											
PLANNED ORDER RELEASES			110		60		15													
PLANNED AVAILABLE INVENTORY																				

Machine Loading Schedule

<u>Load & Capacity</u>

The load refers to the work assigned to a machine or an operator, while capacity is the output volume capable of being produced in any convenient period of time. The department, machine or operator is said to be fully loaded when the load is equal to the capacity. The plant is overloaded if the load exceeds the capacity, whilst if the load is less than the capacity it is underloaded. (Refer to Fig. 7.)

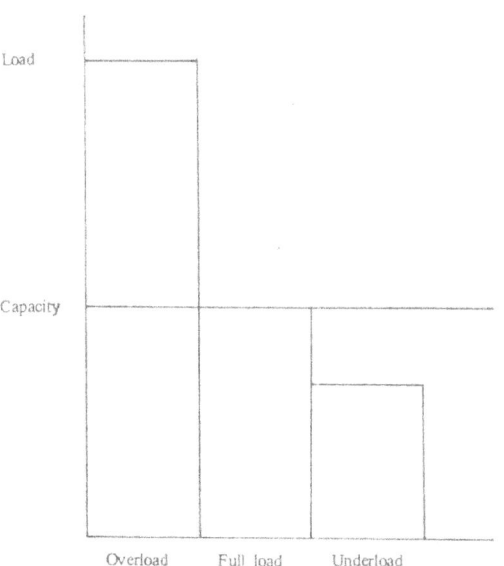

Fig 7

Loading

The preparation of loading schedules for a machine or an operator can provide great improvements in productivity. Ideally, every machine or operator should know the work which has to be carried out for as far forward in time as possible, say, for at least one working week. In this way, it will be possible:

(a) To make maximum possible utilization of plant and personnel.
(b) To set and meet target dates.
(c) To determine whether it is necessary to set up a new plant.
(d) To increase operator morale.

It is basically only possible to express a load in hours of work. However, it is frequently convenient to express it in other physical terms, e.g.:

(a) Weight (a machine has an output of x tons an hour).
(b) Length (a machine has an output of x feet an hour).
(c) Quantity (a machine has an output of x parts an hour).

All these are expressions of standard hours of work at a known rate of working.

The load on a machine/operator specifies the output and time for the period under consideration and may, e.g., take the following form:-

Monday	5.00 am -	5.15 pm	Job 001
Tuesday	8.00 am -	12.30 pm	Job 002
	12.30 pm -	5.15 pm	Job 003
Wednesday	8.00 am -	10.30 am	Job 004
	10.30 am -	12.30 pm	Job 005
Friday	8.00 am -	11.00 am	Job 019
	11.00 am -	5.15 pm	Job 020

This load can only be derived from the schedule, while the schedule in turn can only be formulated after considering the load. As delivery dates determine the schedule, it is necessary to consider the load when establishing a delivery date, this relationship being shown in Fig. 8 below:-

Fig 8

It is foolish to set a delivery date without due regard to the load. Yet this is done so frequently that it can be considered a general practise. Inevitably, this will lead to costly use of labor and to a break in delivery dates and should be discouraged at all costs. When plant capacity is fully utilized it is only possible to insert other work at the cost of existing commitments.

Steps To Take
To prepare the machine loading chart, the following steps should first be taken:-

(a) Collect complete data about machine which is available.
(b) Group similar machines together, i.e., divide all the machines into various groups with respect to the type of work that could be carried out and then subdivide each group with respect to capacity.
(c) After taking due precautions, establish the available machine hours. It should be borne in mind that certain machines, e.g., heavy duty machines, may require more frequent lubrication, while certain machines may be old or less efficient and hence need more maintenance than the rest.
(d) Allocate the total load to the various machines, possibly with the aid of Gantt charts, coupled with economic manufacturing quantities.

MACHINE LOADING CHART

Machine	No. of M/ CS.	Available capacity at 40% per week	1st Week	2nd Week	3rd Week	4th Week	5th Week
10.T Press	1	40	4321				
40.T Press	1	40		4321	4321		
Drills	1	40				4321	
Tapping	1	40					4321

In the above example, we take a look at job No. 4321. Pressing work is to be carried out for 40 hours on a ten-ton press, 80 hours on a 40-ton press, 40 hours on a drill and 40 hours on a tapping machine. The load allocated to each machine for each week is represented by yellow lines while the job No. is indicated above the lines. The achievement line is shown in red superimposed on yellow. Overshooting of the machine load schedule is represented by a blue line. Corrective action should be taken if any portion in the chart is yellow or blue.

A machine may be idle for one of the following reasons:-

(a) Power failure.
(b) Lack of materials.
(c) Lack of manpower.
(d) Lack of orders.
(e) Lack of instructions on what to do next.
(f) Breakdown of machine.
(g) Breakdown of tool.
(h) Waiting for inspection.
(i) Machine being idle from the time it produces the first unit till the part is approved by inspection.

These various breakdowns can be indicated by specific symbols in the machine location chart so that the appropriate departments can take the necessary corrective actions.

The advantages of the machine loading chart are as follows:-

(a) It is possible to forecast delivery as accurately as practically possible.
(b) The unoccupied time of the machine can be filled up by small individual jobs.
(c) Overloading can be eliminated.
(d) Underloading and the presence of excessive machines can be highlighted.
(e) The theoretical parts schedule can be easily translated into a practical works schedule, i.e., implementing the theoretical parts schedule can be easily carried out.
(f) Bottlenecks can be indicated - delay-causing machines can be spotted and corrective action can be taken to overcome the delays which may be due to:

(i) Slow workers.
(ii) Machines which are defective.
(iii) Tools which are defective.

The above are indicated in the chart to facilitate corrective action.

<u>NOTE</u>
In practise, for reasons of economy, a machine has to run ten times its setting up time, e.g., if the setting up time for a job is 50 minutes, the machine has to run for at least 500 minutes. Thus, while scheduling the load for a machine, it should be borne in mind that many different short jobs are not planned for the same machine because for each job the machine has to be set up afresh.

Questions for Review
(a) What are the probable sources of difficulty and how may they be removed when a firm is losing repeated orders from customers because of unreliability in achieving promised delivery dates?
(b) To generate a complete list of parts and subassemblies needed to produce the end item in the required quantities and time-frame, what are the main inputs to the MRP system?
(c) Describe a machine loading system which is appropriate to a factory manufacturing a product of your choice.
(d) What is the relationship between the Master Schedule, MRP/Part Schedule and Machine Loading Schedule?

5 ORDERING/DESPATCHING

Ordering or despatching involves the directing of the different production departments via orders in carrying out the assigned operations as per the manufacturing schedule. These orders are issued by the production planning and control department which is also responsible for following up.

Work Order

A schedule does not give the automatic authority to go ahead and manufacture. The authority to manufacture ("action phase") is provided through the Work Order. This document is also known by these other terms: Shop Order, Factory Order, Job Order, et al.

The Work Order provides critical information on the item (usually work order per item) to be manufactured, e.g., the quantity demanded, material specifications, drawing numbers, machine particulars, routes, starting and completion dates, et al. Fig. 9 is a specimen Work Order form.

The Work Order form is designed for recording information on the actual progress connected with the order. The store(s) or the upstream department in the process route would send work-in-progress items to the department concerned. For instance, Dept. Y in the route XYZ was due to receive 800 units from Dept. X on 6/7/xx and another 400 units the following day, giving a total of 1,200 units.

On 6/7/xx, 800 units were produced by Dept. Y, of which 790 units were accepted by Quality Control while the remaining 10 units were neither accepted by Quality Control nor salvageable. Thus, only 790 units were transferred to Dept. Z. On 7/7/xx, 600 units were produced, making a total of 1,400 units to-date, with 570 units approved by Quality Control. However, the salvage team (assuming there is one in the Quality Control Dept.) salvaged 10 units. Therefore, the total transferred to Dept. Z that day was 580 units, giving a cumulative total of 1,370 units.

Copies of the Work Order should be circulated as follows:-

(a) Copies to all the departments in the route (with the exception of "Department Concerned" information which is department-specific).
(b) One copy for Accounts Dept.
(c) One copy for Production Planning & Control Dept.
(d) One copy for notification of final delivery.

The Work Order needs not be sent to Sales Dept. if the delivery date could be kept; otherwise, only the revised date needs to be communicated.

The Accounts copy provides information for costing. It is necessary to sort out the actual product cost and facilitate the booking of labor and material against the order. On the completion of the work by all the departments/sections, all the copies of the Work Order are assembled in the Accounts Dept. to find out the total expenditure incurred against the sales order.

The Production Planning & Control copy is sent to the planner for the purpose of following up and confirming whether production is according to schedule. Thus, information has to be obtained from the Production Dept. through feedback documents such as reports and delivery tickets. The planner can either input the information about work progress in the respective departments/sections on the back of the Work Order or summarize them on a separate form and upload them on to the Machine Loading Chart.

Fig 9 Work Order

Front

	Work Order	No. _____

Job Description: _____

Job: _____ Qty: _____

Drawing: _____ Sales/Stock Order: _____

Department Concerned: [   ]

Process Route: _____

_____

Date of Start: _____ Order Issued by: _____

Date to Complete: _____ Date of Issue: _____

Remarks:

con't . . .

Back

Receipts From Dept		X			
Date	Qty	Date	Qty	Date	
8/9/XX	800				
	800				
9/9/XX	400				
	1200				

Transfer to Dept		Z			
Date	Qty Made	Qty Passed	Rejected	Salvaged	Total Transferred
8/9/XX	800	780	20		780
	800	780	20		780
9/9/XX	600	560	40	10	570
	1400	1340	60	10	1350

Master Requisition

The Master Requisition is an authorization for the stores to issue the materials on a given Work Order. It indicates to production personnel how much materials are needed for the job and authorizes the latter to draw the materials. The latter prepares an issue requisition for this purpose. A Master Requisition is presented in Fig. 10.

The Production Planning & Control Dept. raises the Master Requisition and sends it to Stock Control for the checking of the availability of materials. After approval of the Master Requisition by Stock Control, the Master Requisition will be sent to the first department/section enroute along with the Work Order. The department/section supervisor will prepare his own requisition, the Material Requisition Form, and send it along with the Master Requisition to the store. When issuing materials, the store will enter the items issued in the Master Requisition.

After each item issued, the cumulative total of issues is noted. Once the full quantity authorized by the Master Requisition has already been issued, the store keeper will refuse further issues. If the supervisor wants more materials, he has to obtain the authorization of his superior, usually the Production Manager, or, the Production Planning & Control Dept.

On job completion, the Master Requisition is sent to the Accounts Dept. for cross checking against the corresponding Material Requisition Forms.

Copies of the Master Requisition should be distributed as follows:-

(a) One copy for Production Dept.
(b) One copy for Production Planning & Control Dept., which is to be filed with the Work Order.

MASTER REQUISITION

DEPARTMENT: _____ NO: _____

SECTION _____ JOB DESCRIPTION _____ DATE: _____

QTY: _____ WORK ORDER NO: _____ AUTHORISED BY: _____

S. No.		Unit	Allocated Quantity	Date Requd	Issue 1		Issue 2		Issue 3		Issue 4		Issue 5	
					Date	Issued / Balanced	Date	Issued / Balanced	Date	Issued / Balanced	Date	Issued / Balanced	Date	Issued / Balanced
	BRASS SCREWS	KG	3500	1/2/XX	2/2	1000 / 1000	4/2	2000 / 3000	5/2	500 / 3500				

Fig. 10 Master Requisition

Final Delivery Notification

Sometimes, after the total units required are delivered by one department to another, the latter department closes its order once the delivery has been made. However, later when a certain quantity of salvaged material is sent to the latter department, the latter department cannot re-open the already closed material account order nor can it utilize the salvaged material in production.

However, for work to be carried out, machines will have to be re-set. Such occurrences can be prevented by ensuring that the receiving department closes an order only after it receives the Final Delivery Notification from the previous department along the route.

The Final Delivery Notification is a copy of the Work Order but different in color. Behind this form, information about the quantity delivered is entered. See Fig. 11.

FINAL DELIVERY NOTIFICATION						
This Work Order is closed in the departments indicated below						
Total Qty Delivered	By	To	Scrap	Salvage	Date Closed	Shop Clerk Sign
10,500	A	B	500	500	11/7/61	

Fig 11. FINAL DELIVERY NOTIFICATION

Questions for Review

(a) In your organization, are there any forms of authorization for production to produce and for store to issue materials?

(b) How is the Ordering/Despatching function related to the other functions of Production Planning & Control?

6 PRODUCTION CONTROL

The decision and the authority to produce should be followed by the below-mentioned:-

(a) Labor utilization is monitored.
(b) Material utilization is monitored.
(c) Production rates (output/time) are as per schedules.

Material Utilization Control
Issue Requisition
On receiving the Master Requisition from Production Planning & Control Dept., the production staff is empowered to draw materials. He prepares an issue requisition, the Materials Requisition, for this purpose. A Materials Requisition form is given in Fig. 12.

The Materials Requisition forms should be distributed to the following departments:-

(a) One copy for the department from where the requisition originated.
(b) One copy for Store & Stock Control.
(c) One copy for Accounts Dept.

As has been mentioned, the store-keeper will issue materials only up to the quantity authorized by the Master Requisition. In this way, material wastage and scrapping are eliminated. If, for any reason, a Work Order cannot be closed unless material is made available, the production staff can draw the material on an excess requisition.

MATERIALS REQUISITION

| Issued On: |
| Issued By: |

No.
Work Order No :
Department:

| | | | | | | | For Accounts Use | |
No.	Description	Code No.	Unit	Quantity Required	Quantity Issued	Remarks (Initials)	Unit Std Cost	Total Std Cost

Requested by _____   Received by _____
Date _____   Date _____

Fig. 12 Materials Requisition

Excess Requisition

For a particular job, the Master Requisition establishes the required quantity of material plus a predetermined percentage for scrap. Sometimes, after drawing this quantity, the supervisor may require more material for production. This additional quantity needed may be due to:

(a) Wrong calculation.
(b) Over-scrapping during manufacturing.

Whatever the reason is, additional material required can only be drawn through the Excess Requisition. If the excess material required is due to wrong calculation on the Process Sheet, the Excess Requisition is charged to the Work Order itself and the Process Sheet will be revised so that the mistake is not made again for future orders.

If the excess material required is due to the error of the Materials Dept., Inspection and/or the Production Dept., the customer should not be made to pay for this mistake. Thus, the excess withdrawal is authorized and charged to the department which committed the error.

The Excess Requisition is usually authorized by the Production Controller and/or the Production Manager, which can be done by stamping the "Excess Approved" certification on a new set (three copies) of Material Requisition forms against which the store will issue the extra materials. The certification stamp is presented in Fig. 13.

Expenditure against Excess Charge figures are collected on a monthly basis and supplied to Management for control purposes.

EXCESS APPROVED
Charge to: Production Controller: Date:

Fig 13 Excess Approved Certificate

Exchange Requisition

Some of the material drawn from the stores may not be up to the standard, i.e., they are defective. For example, out of 1,000 bolts drawn from the store 90 of them do not have proper threads. To complete the work 90 more bolts will ordinarily be charged to the Work Order, i.e., to the customer, if they are drawn on an ordinary requisition.

An exchange requisition system is introduced to overcome this. Through the Exchange Requisition Note the defective material is exchanged. An Exchange Requisition Note is given in Fig. 14.

The Exchange Requisition Note should be distributed as follows:-

(a) One copy for the department from where the requisition originates.
(b) One copy for Store & Stock Control.
(c) One copy for Accounts Dept.

No.

MATERIAL EXCHANGE NOTE

Issued on: Work Order No.:

Issued by: Inspector:

                                       Date

No.	Description	Code Unit No.	Quantity Exchange	Inspector's Remarks	Quantity Issued	Remarks (initials)

Qty. Exchange by: New Qty. Received by:

Date: Date

Fig. 14 MATERIALS EXCHANGE NOTE

Material Return Note

All excess issued materials should be returned to the store so that they do not occupy space on the shop floor and become damaged. To implement this, a Material Return Note is utilized. This document is raised by the production staff, usually the supervisors or foremen. The following can be the reasons for returning materials to the store:-

(a) Excess calculation by Production Planning & Control.
(b) Job cancellation.
(c) Increased labor efficiency.
(d) Excess materials utilized not previously returned.

Usable materials should not be returned to the store as a false means to boost the image of efficiency of the Production Dept. Thus, it is necessary to have the inspector's remarks on the materials returned. Certain materials however may be returned but certified to be reused as secondary quality items. "Cut-off" materials may be returned and reclassified as smaller/shorter standard size pieces. A Material Return Note is presented in Fig. 15.

The Material Return Note should be distributed as follows:-

(a) One copy for Store & Stock Control.
(b) One copy for Accounts Dept.
(c) One copy for the department from where the return originates.

MATERIALS RETURN NOTE

No. ________

Department ____________

Work Order No. ____________

Inspector's Remarks ____________

____

Inspector ________ Date ______

No	Description	Code No	Unit	Quantity Drawn	Quantity Returned	Reasons for Return	For Accounts Use: Unit Std. cost	For Accounts Use: Total Std. Cost

Qty Returned by ____________

Date ____________

Qty Received by ____________

Date ____________

Fig. 15 MATERIALS RETURN NOTE

Labor Utilization Control

Where labor costs are predominant, it is very important to see that labor is fully utilized. In a system of time rate payment, a worker is usually paid for the time he has clocked in irrespective of the amount of work turned out by him. It is important for the management to know whether his time has been utilized efficiently. For this purpose, a Time Booking system, which also helps in arriving at the correct labor costs, can be used.

The Time Sheet is the easiest form of Time Booking adopted.

Time Sheet

In this system, the shop clerk daily records the details of the work done by all the workers on the shop floor in the day. The shop clerk goes round the shop floor collecting data on what work the workers are doing, the amount of time they spend on each item, including breaks or rest, et al. A Time Sheet is presented in Fig. 16.

				No.	
		TIME SHEET			
Department:				Date:	

Operator	Opr No.	From (Time)	To (Time)	Work Order No.	Remarks
C B Tan	121	8.10	9.15	6732	
P K Lian	494	8.15	12.05	6732	
P T Han	212	8.05	8.30	-	Dispensary
T T Lam	155	8.20	?	6730	

Fig. 16 TIME SHEET

Though the system and the form used are very simple, it develops much difficulties in implementation. The shop clerk has difficulty in knowing when a worker changes his job. Should the shop clerk fall behind schedule in preparing the Time Sheet, he usually asks the worker what tasks he has been performing and gets only approximate data at best. He is thus not in a position to compare performance with standards.

For control purpose, it is important that the times booked are accurate, i.e., the shop clerk should check whether the worker has been producing the quantity expected in the time he is supposed to have worked.

There are a number of reasons why the system may fail to provide the correct information, e.g., obtaining accurate information from 50 or more workers will be a difficult task for the shop clerk. This difficulty could be eliminated by utilizing the Daily Operator's Report.

Daily Operator's Report
The recording of timings could be entrusted to the worker himself in order to obtain accurate timings. The timings will be recorded in the Daily Operator's Report. Should the worker be found to have recorded false timings, the shop clerk would fill in the work outturn and the standard time for this outturn. The Daily Operator's Report is presented in Fig. 17.

DAILY OPERATOR'S REPORT

| Operator: | C T Wong | | | Opr No.: 134 | | |
| Department: | Polishing | | | Date: | 5/2/82 | |

Work Order/ C. T. No.	Operation No.	From	To	Elapsed Time	Quantity Produced	Time Allowed
4321	Grinding	7	9	2	81	2.5
C.T. 29		9	10	1		
4321	Inserting	10	12	2	90	3.5
4321	Pin-Cutting	1	2	1	50	1.5
C.T. 31		2	4	1		
4325	Pin-Polishing	3	4	1	50	2

Fig. 17 DAILY OPERATOR'S REPORT

In the Daily Operator's Report, the last column will be filled by the shop clerk, while the worker will fill in the other columns.

When the Daily Operator's Report is completed, it will be verified by the Supervisor/Foreman before being passed to the Production Planning & Control Dept. and the Accounts Dept. for recording the progress of production and for costing the jobs respectively.

In the Time Sheet and the Daily Operator's Report, C. T. Nos. indicate certain expenditures chargeable to overheads such as workers going to the dispensary or being idle due to lack of materials, et al. The Time Sheet and Daily Operator's Report provide details of the time spent on non-productive activities as well. This form of time-keeping is suitable for small and medium sized factories.

Job Card
In big organizations, a Time Recorder is normally provided instead of having the workers manually recording the time; the worker inserts a card into the clock and the time is automatically recorded on the card. The worker punches the card on and off for each Work Order and C. T. No.

To help the worker to position the card correctly without overprinting, a clip-type Time Recorder is used. The clock is used in this Time Recorder which clips off a small portion of the card so that the card will be automatically positioned for the next clocking. The Time Recorder avoids overprinting because of its clip-off nature. A Job Card used in this system is presented in Fig. 18.

JOB CARD				Date	Qty Produced	Time Allowed	Time Booked	Difference
Name: Ibrahim Staff No.: 271 Dept: 21. M.W.E. on 20/11/61								
W.O	On (Time)	Off (Time)	Elapsed Time					
4321	Mon 7 am							
		Mon 12.00	5					
C.T.	29 Mon 1 pm							
		Mon	1					

FRONT   BACK

Fig 18. JOB CARD

On the reverse of the card, the production time of the worker and the standard time are linked to one another.

For large factories, one Job Card Time Recorder can record 200 timings simultaneously. One master clock should be centrally located to control the time of the "slave" clocks in the various departments in order to ensure accuracy.

Production Progress Control
The difficulties involved in production scheduling will be minimized if all scheduled work is performed as planned. This is however an unlikely scenario because of machine break-downs, material shortages or out-of-stock situations and other unforeseen circumstances. Therefore, production efficiency will not always be up to expectation.

Good records of production progress should be maintained so that revised schedules and material re-provisions can be made when necessary to minimize costs and provide good customer service. A production planning and control system which is otherwise well designed could turn out to be ineffective due to inadequate production progress reports and records.

Timely, accurate and adequate information related to production progress is required by the Production Planning & Control Dept. and the Accounts Dept. Since the information used by these two departments is similar in many ways, one reporting system could be designed to serve their needs. The following are some important basic information required by both departments:-

Progress Reporting Information	Information Used By PP&C Dept	Accounts Dept
Work Order Number	X	X
Operation	X	X
Date and Time or Report	X	X
Quantity Produced	X	X
Quantity Scrapped	X	X
Production Time	X	X
Explanation of Time Variances		X
Operator Name and/or Number		X

The Job Card and Move Cards are usually utilized to capture this information.

Job Cards
The completed Job Cards, which have been mentioned earlier, are forwarded to the Production Planning & Control Dept. for the latter to post the progress of production, and, to the Industrial Engineering Dept. and the Accounts Dept. as a basis for gauging the efficiency of the shop-floor and the actual costs incurred.

Move Card/Partial Deliver Ticket
The Move Card or Partial Delivery Ticket is utilized to assist the Production Planning & Control Dept. in keeping track of the work-in-process of each job through the Production Dept. and in expediting and updating the production schedules.

It represents the authorization to move batches of products from section to section or department to department along the process route.

The Move Card or Partial Delivery Ticket is usually raised by the delivery section/department, countersigned by the receiving section/department, and retained by the delivering section/department. At the end of each shift, all the Move Cards or Partial Delivery Tickets are collected and submitted to the Production Planning & Control Dept. by the Supervisor/Foreman. The Production Planning & Control Dept. later submits it to the Accounts Dept. A Move Card is presented in Fig. 19.

MOVECARD/PARTIAL DELIVERY TICKET

Job Description: Date:
Work Order No.:   Shift:

	Section	Section Foreman	Remarks
From			
To			

Quantity Delivered:

Fig. 19 MOVECARD/PARTIAL DELIVERY TICKET

Daily Production Report

The Daily Production Report is utilized to further enhance the control of production progress. It reports the daily output of each department and the time booked for each operation within the department.

The input for the preparation of this report is obtained from the Job Cards of the Daily Operator's Report and the Move Cards or Partial Delivery Tickets.

At the end of each shift or work period, the report is forwarded to the Production Planning & Control Dept. who will utilize it in the maintenance of the Production Schedule. A Daily Production Report is presented in Fig. 20.

Immaterial of the type of production progress record, the record should be carefully studied and analysed to determine whether there is a need for rescheduling or other corrective action to rectify the situation before excess costs are incurred or poor customer service arises.

DAILY PRODUCTION REPORT   Date _____

Department _____
Shift _____

No.	Job Description	Work Order No.	Operation	Booked Time	Units Processed	Units Pass	Units Rejected	Units Salvaged	Remarks
						Prepared by	_____		
						Date	_____		

Fig. 2/ DAILY PRODUCTION REPORT

Questions for Review

(a) In the controlling of material utilization, what are the documents used?

(b) Discuss whether excess issued materials should accumulate on the shop-floor till the end of the week before being returned to the store or be returned immediately to the store.

(c) Compare and contrast the Time Sheet, the Daily Operator's Report and the Job Card used for the recording and controlling of labor time utilization.

(d) If production is behind schedule, what should be done?

7 MANUFACTURING RESOURCE PLANNING: MRP II

In Chapter 4, MRP has been brought up. MRP has now expanded to mean more than material requirements planning. There is now have MRP II, which represents manufacturing resource planning, a system for planning and controlling the operational, engineering and financial resources of a manufacturing firm.

Classes of MRP Users
Oliver Wight in his book on manufacturing resource planning defines four classes of MRP users. The characteristics of the four classes are listed in the table below. The lowest level is Class D, in which the potential utility of material requirements planning is hardly realized. MRP is basically used as a data processing system with many of the traditional production control procedures (e.g., expediting and shortage lists) still being used. The Class A MRP user is at the top of the list. This class of user is a company that uses material requirements planning together with capacity planning, shop floor control and other components of a computer-integrated production management system (CIPMS), which as the name implies, involves the use of the computer. It is a powerful tool for helping to accomplish the vast data processing and routine decision-making chores in production planning which had previously been carried out by human beings. MRP II is the next step beyond the Class A user. To describe MRP II, let us first examine the progressive evolution of material requirements planning into manufacturing resource planning.

Classes of User and their Characteristics
Class A
(a) Uses closed-loop MRP.
(b) Integrated system has MRP, capacity planning, shop floor control, vendor scheduling, et al.
(c) MRP system used to help plan sales, engineering, production, purchasing, et al.
(d) No shortage lists to override the production schedules.

Class B
(a) System has MRP, capacity planning and shop floor control, but no vendor scheduling.
(b) Not much used in managing the business - it is used as a production control system.
(c) Needs help from shortage list.
(d) Inventory is higher than needed to be.

Class C
(a) System is used for inventory ordering rather than scheduling.
(b) Scheduling by shortage list.

(c) Master schedule is overloaded.

Class D
(a) MRP works in the data processing department only.
(b) Inventory records are poor.
(c) Master schedule, if it exists, is overstated and mismanaged.
(d) Relies on shortage list and expediting rather than MRP.

The Four Steps of MRP
Over the years material requirements planning has changed significantly. The four steps below can be identified in the evolution of MRP:-

(a) *Improved ordering method.*
(b) *Priority planning.*
(c) *Closed-loop MRP.*
(d) *MRP II.*

The first step was implemented when initial use of the computer was made to perform the calculations of requirements planning. Before the computer era, this task was performed manually and took up a great amount of time and manpower. Computerized MRP systems represent a vast improvement in the ordering of raw materials and components because of the speed and accuracy with which the requirements planning task could be carried out.

The need for the second step in MRP evolution arose out of attempts to implement step one MRP in conjunction with an unrealistic master schedule. This was a master schedule which ignored the limitation imposed by plant capacity and other constraints. It resulted in the MRP processor generating schedules and requirements which could not be accomplished by the factory. The use of shortage lists continued, as a result. The MRP system began to incorporate priority planning into their computations in order to overcome these problems. The term "priority planning" not only connotes an MRP system which determines what materials will be required, it also means that the planning of material requirements can be phased into time periods (weeks, or even days). Priority planning, besides providing a means for dealing with rush jobs by increasing their priorities, also helps to un-expedite jobs whose priorities have been reduced.

The level of achievement of the Class A MRP user is represented by step three MRP. Closed-loop MRP represents an improvement over step two MRP as it not only plans the priorities but provides feedback information related to executing the priority plan as well. In closed-loop MRP the various functions in production planning and control (inventory management, capacity planning, shop floor control and MRP) are integrated into a single system. There is also feedback from vendors, the production floor, et al., when there are problems in implementing the production plan.

MRP II

Closed-loop MRP is a great achievement in terms of bringing together the various separate functions of a production planning and control system. MRP II represents the last step in the evolution of MRP (at least, as it has been conceived). This fourth step involves a tie-up between the closed-loop MRP system and the financial system of the organization. This combination is known as manufacturing resource planning.

MRP II has the following two basic characteristics which extend beyond the closed-loop MRP:-

(a) It is an operational and financial system.
(b) It is a simulator.

Being an operational and financial system makes MRP II a company-wide system, involved in all aspects of the business, including sales, inventories, engineering, production and cash flows. The operations of the various departments are reduced to the same common denominator, namely, financial data. This common base provides the management of the company with the information required to manage the company successfully, e.g., raw materials on hand can be converted into their equivalent cost and summed over all stocks in inventory, work-in-process can be evaluated by adding raw material costs to the cost of labor turned in against the particular part numbers and orders, and, other data relating to operations can be expressed in monetary terms by a similar method of calculation.

MRP II is also a simulator which is aimed at answering "what if" questions. It can be utilized to simulate the likely outcomes of alternate production plans and management decisions that are under consideration.

Manufacturing resource planning is essentially quite similar to a computer-integrated production management system (CIPMS). The latter includes not only the operational system (inventory management, capacity planning, MRP, et al.), but also the cost planning and control module which is linked to the company's accounting and financial systems.

Questions for Review

(a) What are the classes of MRP users?

(b) Describe the basic characteristics of MRP II.

8 THE PRODUCTION PLANNER AND HIS PROBLEMS

As mentioned earlier on, the production planning department is the nerve center of a factory. The production planner keeps in close touch with the staff of the production floor, monitors the progress of production and reschedules the production as and when necessary. His job is not an easy one. He has to keep track of many things all at once. If he forgets or misses a few important things, chaos could result. For example, if he forgets to instruct the die-maker to prepare dies for a new model scheduled for production soon, the new model would not be produced on time. The production planner has to be methodical in his work, he has to have a head for details if he is to perform well and remain calm under pressure. Pressure there is bound to be, what with last minute orders from customers, delays in production work due to machine break-down or power failure or even labor shortage, shortage of raw materials, and other unforeseen disruptions.

Functions

The production planner normally performs several important functions. One of these is inventory control. He estimates the material requirements at any one time and ensures that there is sufficient stock of materials and no overstocking which would result in higher storage and material handling costs. Knowledge of inventory control methods such as JIT (Just In Time) might be useful. If he were involved in MRP or MRP II, he has to have some computer knowledge. He also has to liaise with the purchasers regarding what, when and how much materials to buy.

Secondly, the planner monitors the progress of production, e.g., the status of work in process, finished goods, and so on. He has to reschedule production work when production targets are not met.

Thirdly, there comes the actual work of production scheduling itself. He has to have knowledge of the production flow (probably, if he is new to the job, he needs to consult a production flow chart or even a quality control flow chart). He has to have product/technical knowledge. He might already have acquired some technical knowledge from college or university but in most cases, he would still require on-the-job training, especially on the technical aspects. He might be expected to understand blue-prints or schematics. He has to have an aptitude for figures as his work involves mainly figures.

Fourthly, though this is basically the task of the sales department, he might have to prepare sales forecasts. He might even have to prepare the master schedule for the whole factory, though this is usually carried out by a more senior person, such as the production control manager. Of course his basic job is to schedule production for each machine or section.

Fifthly, he schedules deliveries to customers, ensuring the right quantities and the right items are delivered according to schedule. He might even have to liaise with customers regarding such matters. For the task of scheduling, he has to have data on time standards. In other words, he has to know the time it takes to perform the various sub-operations, and, of course, the time it takes to perform a complete operation. Information of time taken to perform various jobs is normally gathered by work study officers or industrial engineers, who carry out what we call "work measurement" and give advice on methods improvement. The production planner relies on his judgment and experience when scheduling production. At times, he has to consult the line supervisors directly responsible for production to find out whether schedules are realistic and achievable. Scheduling is very much a "numbers game". If the planner detests figures, production planning is not the work for him. Depending on how complicated the product-to-be-produced is, the task of scheduling requires patience, an eye for details, and thoroughness. As production normally has to proceed without delay or stoppage, the production planner has to work fast, to reschedule production which has fallen behind schedule, or to advice on "change of items" as when there is a sudden cancellation or delay of orders or delivery. Technical knowledge and knowledge of production processes are essential.

Problems and Dos of Production Scheduling
The purpose of a production plan is to balance requirements (orders in hand/sales forecasts) against resources (people, materials, machines) so that the company operates at maximum efficiency and profitability.

The first step required when formulating a production schedule is sorting the orders or forecasts received into a sequence, according to the dates by which they are required.

The orders have then to be broken down to establish what components need to be bought or made to complete the order. From this, the production planner can determine what material, machines and labor hours will be needed and the earliest possible date for final production. This is a crucial process as, although there may be standard lead times, circumstances may change from day to day, even from hour to hour, e.g., being short of just one component can delay the whole order, leaving the factory with an excessive level of unfinished work in progress.

It is essential to consider the total factory loading, especially if the company makes a wide range of products. Unless each line is loaded evenly, one part of the factory may be working flat out, paying premium rates for overtime, while in other areas people and machines may be standing idle. In most manufacturing processes there is a period of non-productive time when orders are changed over. Requirements for the same or similar parts should be batched together to minimize downtime and to achieve greater output. This balancing process should result in a schedule which can produce goods at a profit and satisfy the customer by the efficient use of available resources. There will be times when orders either cannot be produced on time or only at an unacceptable cost. In such cases the customers should be informed. It is better to lose a single order and keep a good customer than to gain a reputation for unreliable delivery promises and thus lose customers in the long run. Achieving the best balance of resources while at the same time satisfying customer demands is a far from easy task.

In any production process, it is usually impossible to state precisely what can be produced with the given equipment, manpower and materials. Only the industrial giants can afford computer power to formulate the best production plan, according to the wide range of possible resource allocation against a sequenced order plan. Most companies have to rely on simpler systems, coupled with the flair and experience of their production planners. As computerization becomes cheaper and more powerful, companies should regularly consider whether they can benefit from the mini- and micro-based systems appearing in the market.

Production scheduling is simpler if machinery is highly specialized rather than multi-purpose. Within the priorities of the order plan, production planners should ensure that all machines are loaded to the maximum possible and that downtime (due to, e.g., order changeover or gaps in the production sequence) is kept to the minimum.

When planning material availability, the production planner should ask the following questions:-

(1) What material is required and when?
(2) Is the material available in stock at the moment?
(3) Can the material be made available in time for the planned start of production at an acceptable cost?

Making use of available labor to the maximum is a complex factor which the production planner has to take into consideration. Particular equipment may require specialist skills which only a few workers possess. There may be a long holiday period. It may be a time of the year when absence through sickness is traditionally high. The work-force may also resist changes in shift patterns or levels of overtime working. There may even be serious labor shortage. Industrial health may be poor resulting in "go-slows" or strikes.

To overcome such problems it may be necessary to subcontract work out. Other companies operating in the same industry may have spare machines and manpower capacity, or may have the right materials available. Reversing the process, i.e., taking on subcontracted work, can also be a useful way of levelling out workload troughs. Overtime work may be another solution.

Balancing all the variations in resources against demand should result in a good production plan which would see to it that customer orders are fulfilled on time. It has to be borne in mind here that production scheduling is an individualistic task which depends on the planner's own experience, judgment and creativity. Different production planners schedule a similar task differently, i.e., no two production plans are alike.

Production Control

A production plan is rarely, if ever, carried out in all its details. Machines break down, suppliers do not deliver on time, workers fall sick, take holidays, leave the company or go on strike. It is the task of the production controller to ensure that production is maintained in line with the production plan wherever possible. The production controller has to respond to the things which do go wrong and rework the plan in order to get back on schedule.

The production controller monitors the supply and production process to ensure that there is, at all times, accurate up-to-date information on what performance has been achieved and what actions are required to maintain performance. Where deviations from the production plan are spotted, corrective action should be taken at an early stage to overcome the shortfalls. The production controller should make regular checks on material availability and not just on the day it is due. This is because a supplier's promise given some time ago may not be kept. Such routine follow-ups should be backed by regular stock checks.

The production controller should be constantly aware of what labor hours are available, how the operation is performing in terms of quality and productivity, and even what social events are taking place, e.g., planned overtime work may be suddenly cancelled when a big sporting event is unexpectedly rescheduled.

If machines break down, outside contractors can be brought in to repair them quickly. On the other hand, machines can be hired or the job may be subcontracted for completion. Materials not available from one source can be bought from other sources. Alternatively, other materials can be substituted. Defects can sometimes be rectified. Design engineers may agree to the use of off-standard components if they do not materially alter the finished product. Manpower problems may be resolved by resorting to extra overtime work, additional shifts and engaging temporary labor.

Customers will often accept a later delivery date, or accept part shipment of an order, if treated correctly. The production controller should always make the customers aware of the status of their orders.

The function of production control should be the continuous process of checking, reworking and rescheduling. The production controller should see to it that the original production plan is adhered to as closely as possible, though this is vastly difficult. Changes should only be made if they are really unavoidable.

Important Elements Of An Order Plan
(1) Sequencing orders according to dates desired by customers.
(2) Breaking down orders into details.
(3) Maintaining balance between product lines.
(4) Grouping orders together so that there are economies of scale.
(5) Maintaining a balance between "high profit" and "low profit" orders.
(6) Informing the customer of delays.

Salient Points Of Production Scheduling
(1) Planning of materials according to the priority of the order.
(2) Minimizing of machine break-downs.
(3) Accurate recording of stock levels.
(4) Maximum usage of available manpower.
(5) Subcontracting out for the purpose of smoothing workload peaks and troughs.

Pointers On How To Balance Resources And Customer Needs In A Production Plan
(1) Look at orders from customers.
(2) Look at sales forecasts.
(3) Look at stock level or sub-assemblies.
(4) Consider the details.
(5) Plan according to dates required.
(6) Have customer satisfaction in mind.
(7) Identify special requirements.
(8) Aim for even distribution of work-load.
(9) Plan for batches of similar groups.
(10) Consider the machinery required.
(11) Consider whether manpower is sufficient and whether the required skills are available.
(12) Consider subcontracting of work where necessary.
(13) Consider the materials required, when they are required, whether the stock level is sufficient, and whether they can be procured in time.
(14) Aim for a balance of "high profit" and "low profit" orders.
(15) Consider effect of premium payments on profitability.
(16) Look into downtime costs from excessive change overs.
(17) Look into the cost of subcontracting work.

Problems That Can Affect Production

(1) Production quality that is substandard.
(2) Low productivity.
(3) Machine breakdowns.
(4) Unforeseen raw material shortages.
(5) Inferior products, which are rejected by customers.
(6) Strikes at suppliers' plants which affect delivery of materials.
(7) Breakdowns in transport.
(8) Bans on overtime work.
(9) High absentee rate.
(10) Substandard raw materials.

Possible Solutions To Production Problems

(1) Rework substandard products, effect changes in engineering design.
(2) Implement productivity bonus schemes or incentive schemes.
(3) Enter into machine servicing contracts. Plan and implement preventive maintenance. Hire machines.
(4) Use substitute raw materials. Order materials from another supplier who can supply them. Modify or rework some of the parts.
(5) Modify or rework product. Change its design. Give customers improved products.
(6) Keep buffer stocks. Use other suppliers.
(7) Use alternative modes of transport, e.g., air instead of sea.
(8) Recruit more workers. Use temporary labor. Improve labor relations.
(9) Get workers to work overtime. Introduce additional shifts. Improve industrial relations.
(10) Complain to suppliers about substandard quality and ask for improvement and/or replacement. Alternatively, change suppliers.

CASE STUDIES
Case No. 1
In early 1982, JCB, the UK earth-moving equipment manufacturer, was faced with a total shut-down of their highest volume production line as a result of a protracted strike at the factory of their only engine supplier. If you had been the manager of JCB, what would you have done to overcome the problem of lack of material? [Solution is at the next page.]

Case No. 2
Ford Motor Company in Dagenham, England, which was at the time producing 1,000 vehicles a day, was threatened with a complete stoppage of work at their Cortina assembly line, due to a shortage of steering wheels which prevented vehicles being driven off the line. As manager of Ford Motor Company, what line of action would you have taken? [Solution is at the next page.]

SOLUTIONS
Solution to Case No. 1
JCB resorted to the buffer stock they had held for cases of emergency. They also put in a crash engineering development plan to redesign the unit to use an engine from an alternative supplier. The quick redesign combined with an extensive material procurement exercise enabled JCB to fit the new engine and thus maintain the production of their vehicles.

Solution to Case No. 2
A squad of mechanics was used to drive the cars into the park and remove the steering wheels. The steering wheels were then used as "slaves" to drive off more cars. In this way, production was able to go on without stopping. Fitting new steering wheels when they became available was a relatively simple exercise.

Examples of what can upset production.

Questions for Review
(a) Discuss the functions of the production planning department.
(b) Explain how you would schedule production.
(c) What are some of the problems faced by the production planner? How would you resolve them?
(d) Why is production control important?

9 SUMMARY AND REVIEW

Production planning and control represents the heart of an organization. With the production process, materials are transformed into products through the use of men, machines and equipment. In order that products of the required quality and quantity are produced in the required time at the optimum cost, all these production factors have to be planned, organized, coordinated and controlled. This will lead to higher productivity and profitability in the organization.

An effective production planning and control system can help to achieve higher productivity and profitability through forecasting, preplanning, scheduling, ordering and controlling all aspects of production. It can minimize delays, reduce unproductive work, avoid last minute rush work, and, improve working relations and morale in the organization, as a result.

Highlights
The three key areas of a production planning and control system are production scheduling, ordering and production control. The design of the forms used however is also important for the effective implementation of the production planning and control system.

Action Guidelines
(a) Study the current system of production planning and control in your organization.
(b) Analyse/evaluate the present procedure and document flow. Also abide by the following
     principles:-

   (i)   All activities should be necessary and productive.
   (ii) All activities should be as simple as possible.
   (iii) All processes should be arranged in such a manner that a smooth flow results.

(c) Design better procedure based on the material used.
(d) Test the improved procedure.
(e) Implement the improved procedure.